Developing **literacy** Skills

Through Design & Technology

KEY STAGE 1: Y1–2

Kate Faircliffe

Hopscotch

A division of MA Education Ltd

Hopscotch

A division of MA Education Ltd

Published by Hopscotch,
a division of MA Education,
St Jude's Church, Dulwich Road,
London, SE24 0PB
www.hopscotchbooks.com
020 7738 5454

©2012 MA Education Ltd

Written by Kate Faircliffe

Designed by Claire White,
Fonthill Creative, 01722 717029

Illustrated by Kerry Bailey

ISBN 978 1 90751 544 6

Contents

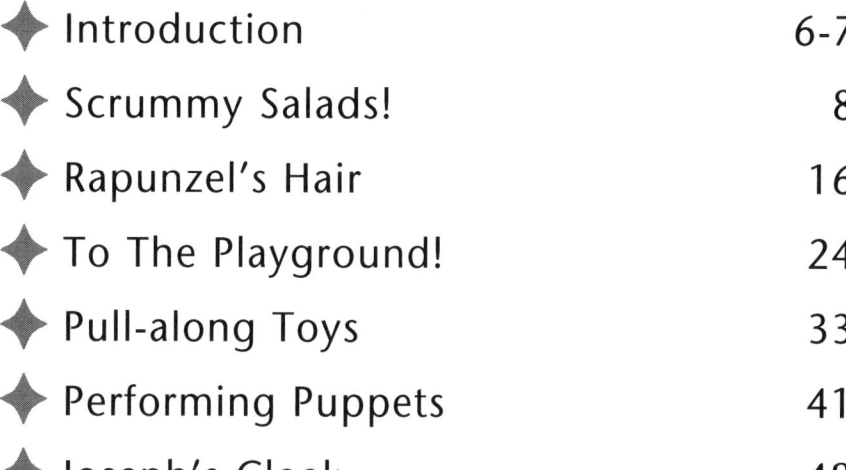

Literacy through design & technology

Introduction

  ABOUT THE SERIES

Developing Literacy Skills Through Design & Technology is a series of books aimed at developing key literacy skills through a range of DT projects, for Key Stage 1 and 2.

It offers a structured approach, providing detailed lessons plans to teach specific literacy and DT skills. A unique feature of the series is the provision of differentiated activities aimed at reducing teacher preparation time. Suggestions for follow-up activities for both literacy and DT ensure maximum use of this resource.

ABOUT THIS BOOK

This book is for teachers of children at Key Stage 1. It aims to:

+ Develop children's literacy and DT skills through a series of stimulating and engaging projects, with supportive and challenging differentiated activities
+ Support teachers by providing practical teaching approaches, based on whole class, group, paired and individual teaching
+ Encourage the development and planning of creative ideas, as well as building on key skills of evaluation and communication

 **CHAPTER CONTENT**

 Literacy objectives

These outline the aims for the literacy activities suggested in the teaching unit

DT objectives

These outline the aims of the DT objectives covered in the teaching unit

 Resources

This lists the resources needed to teach the unit

 Starting point: Whole class

This provides ideas and some key questions with which to get the unit started

 Using the photocopiable text

This explains how to use the texts provided within the unit

 Group activities

This explains how to use the differentiated sheets once the children have split into group work

 Plenary session

This suggests whole class activities to aid the discussion of learning outcomes and future work

 Follow-up ideas for literacy

This suggests further literacy activities related to the teaching unit, to be taught in separate lessons

Follow up ideas for DT

This contains suggestions for further DT activities which could be taught separately

- www.curriculum.qcda.gov.uk/
  The National Curriculum programmes of study and attainment targets for key stages 1 - 4

- www.data.org.uk
  The Design and Technology Association which represents those involved in design and technology education

- www.ngflnorthumberland.co.uk
  Free online resources and curriculum materials for teachers

- www.nuffieldfoundation.org/teachers
  Useful educational links for primary design and technology

- www.nationalschoolspartnership.com
- A website providing educational and teacher resources

- www.techitoutuk.com
  Links, materials and design and technology projects

- www.mechanical-toys.com
  Website detailing the history of mechanical toys, how to make them and how the mechanisms work

- www.cabaret.co.uk
  Museum of automata, with numerous models to make and inspire

- www.robertsabuda.com
  Ideas and templates for pop-up books

- www.footwearhistory.com
  Website detailing the history of shoe design and construction

- www.mechanicalmonkey.co.uk
  Source of different mechanisms and kits

- www.bakerross.co.uk
  General all-purpose class kits for all areas of the curriculum

- www.tts-group.co.uk
  Class resource kits for design and technology projects

- www.spartacus.schoolnet.co.uk
  A website providing free access to a wide range of design and technology materials, resources and software for students to use as they engage in design and technology activities as part of the UK National Curriculum.

- www.teachable.net
  Teacher-developed resources that are adaptable and easy to use

Chapter 1

Scrummy Salads!

 Aim

To design and make a bowl of fruit salad to a particular specification

 Time Allocation

4 x 60 minute sessions

 Literacy objectives

- ✦ To make collections of descriptive words
- ✦ To make simple lists

 DT Objectives

- ✦ To know there are a wide variety of fruit and vegetables that can be named and grouped
- ✦ To learn about health and safety issues when preparing and tasting food

 Resources

- ✦ Scrummy Salads! worksheets. These can later be collated into a booklet, with a front cover designed by the children, possibly incorporating a photograph of their individual salad
- ✦ A selection of fruit, some familiar to the children and some which are not
- ✦ Necessary quantities of all ingredients
- ✦ Tools and equipment e.g. knives, spoons, mixing bowls, chopping boards etc.
- ✦ Plastic table covers, antibacterial cleaner, hand washing and washing up facilities, aprons etc.

 Starting point: Whole class

Start off by identifying fruit with children through descriptions using all of the 5 senses. Ask the children to guess which fruit is coming out of your shopping bag by listening to clues e.g. 'I'm holding a fruit which is round', 'The skin is bumpy', 'This fruit is very juicy', 'The colour is orange'. (It is obviously important prior to this whole activity to find out if there are any children who are sensitive to particular foods, and to adapt the unit accordingly.) Write up the names of the fruits as they come out of the bag and at the end canvas the children on which fruits they like, recording their responses with tallies. This information could later be used to produce a simple pictogram.

 Group activities

Split the children into groups to complete a tasting test of various fruit. This session could also be followed by an art lesson, drawing and painting the fruit from still life, to use in display work with the descriptive words the children have thought of.

 Using the Activity sheets

Activity sheet 1: this is for pupils who will need support reading the vocabulary and completing their sheet.

Activity sheet 2: this is for pupils who will need some support with the reading and writing of the descriptive vocabulary.

Activity sheet 3: this is for pupils who will be able to independently complete the task, thinking and writing their own descriptive vocabulary, referring back to the previous whole-class discussion for ideas if necessary.

Reviewing their thoughts from the tasting task, tell the children that they are going to work in groups to create a fruit salad for another group of children in the class. The first thing they will need to do is conduct a questionnaire, to discover what ingredients they will need in order to make a fruit salad liked by the group. Spaces have been left on the fruit chart, in order for children to customize this to reflect the fruit on offer.

Following the completion of the questionnaire, each group should decide which fruits they will use in their fruit salad, and what, if any, other accompaniments they may include. Explain to the children that they are going to use their findings to complete a 'Fruit Salad Specification' – a plan, which will help them make the best fruit salad they can. They will be working as a team, so will need to come to agreed joint decisions. Remind the children how to make simple lists for their ingredients before the groups split off.

Scrummy Salads!

 ### Developing their designs

Having decided on the ingredients for their fruit salad, spend some time talking to the class about the health and safety issues of working with kitchen implements. Discuss the risks with using knives and spoons for cutting and peeling, and the importance of hygiene when handling food.

Demonstrate to the children different ways of presenting the fruit – size and shape of cut pieces etc. Allow time for changes/additions to be made to their specifications prior to the practical work to reflect this discussion.

Once it is time for the practical activity of making their fruit salads, remind the children to use their specification sheet, remember all the pointers about safe and hygienic food preparation and ensure that they share out the tasks fairly in their group. Each group should also take responsibility for clearing away properly at the end.

 ### Plenary session

When each group has made their fruit salad, they can fill in the top part of the Evaluation sheet. Salads can then be given to their target group to try, and the bottom half of the Evaluation sheet can be completed. Prior to the salads being eaten, a photograph could be taken to use for display work later.

Gather the children back into one group for a final class evaluation, sharing the work done by the children. Use the questions on the Evaluation sheet to prompt discussion and reflection on the task.

 ### Follow-up ideas for Literacy

✦ Read stories and poems on the theme of 'Fruit'
✦ Alphabetical work, e.g. sort different foods into alphabetical order
✦ Complete labelled diagrams of their fruit salads
✦ Compile a class fruit salad recipe book

 ### Follow-up ideas for DT

✦ Challenge the children to design fruit salads for a particular season, e.g. Winter Salad
✦ Extend work done with food preparation (cutting, grating etc), food measurement and planning an order of work
✦ Link this activity with a seasonal display and make papier mâché models of different fruits/vegetables, e.g. Harvest

◆ Tasting Fruit! ◆

✦ For each fruit you taste, draw and describe it, using these words to help you

  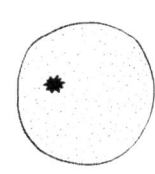   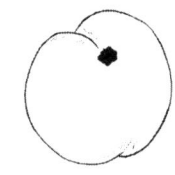

Red   Green   Yellow   Orange   Blue   Soft   Hard   Round   Large/small
Squishy   Hairy   Rough   Bumpy   Silky   Smooth   Crisp   Sweet   Juicy   Fresh

Draw the fruit	Describe the fruit	☺☺☹

✦ Tasting Fruit! ✦

✦ **Use the words below to help you describe each fruit you taste**

  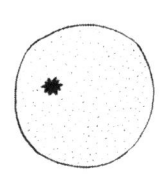   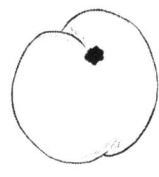

Red   Green   Yellow   Orange   Blue   Soft   Hard   Round   Large/small
Squishy   Hairy   Rough   Bumpy   Silky   Smooth   Crisp   Sweet   Juicy   Fresh

Fruit	It feels...	It looks...	It tastes...	☺ 😐 ☹
Peach	Soft, silky	Red, soft, round	Sweet, like honey	

Name _____

✦ Tasting Fruit! ✦

✦ Taste a piece of each of the fruits on your table and fill in the table below.

  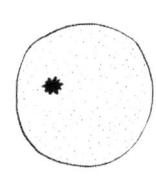   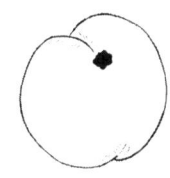

Fruit	It feels...	It looks...	It smells...	It tastes...	It sounds...	☺☺☹
Peach	Soft, silky	Red, soft, round	Sweet	Sweet, like honey	Juicy	

Name _____

  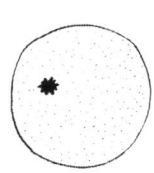   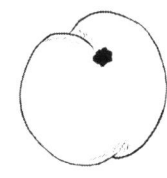

People questioned:

✦ **Choose your 4 favourite fruits from this list:**

Fruit name	Tally marks

✦ **Please choose your preference:**

	Yes	No
Fruit peeled?		
Pips removed?		
Sugar added?		
Fruit juice added?		

My Fruit Salad Specification

We will use these fruits:

◆ We will/will not peel the fruit   ◆ We will/will not add any sugar

◆ We will/will not take out any pips   ◆ We will/will not add any fruit juice

◆ Our fruit salad should taste ....

◆ Our fruit salad should feel ....

Our fruit salad should look like:

The shape and size of our 4 fruits will look like this:

Evaluation

Evaluating My Fruit Salad	Yes	No
Did my fruit salad include 4 fruits?		
Did my fruit salad taste as I wanted it to?		
Did my fruit salad feel as I wanted it to?		
Did my fruit salad look as I wanted it to?		

✦ What was difficult about this task?..

....

✦ What was easy about this task?..

....

✦ Would I do anything differently another time?..

....

Evaluating Your Fruit Salad	Yes	No
Did your fruit salad include 4 fruits?		
Did your fruit salad taste good?		
Did your fruit salad feel nice?		
Did your fruit salad look good?		

Rapunzel's Hair

 Aim

To make a simple moving picture

 Time Allocation

5 x 60 minute sessions

 ## Literacy Objectives

+ To write a simple non-chronological report
+ To link with work done on traditional stories and fairy tales

 ## DT Objectives

+ To design a product with a particular purpose
+ To evaluate what they have done

 ## Resources

+ Rapunzel's Hair worksheets. These can later be collated into a booklet, with a front cover designed by the children
+ Storybook 'Rapunzel' and some examples of books with moving parts (to include a variety of mechanisms)
+ Resources with which to make moving pictures, e.g. cards, split pins, scissors, colouring pens/pencils etc.

 ## Starting point: Whole Class

Read the story of Rapunzel to the whole class and discuss the story. Focus then on the pictures and discuss what is in each picture, what's happening, what's about to happen etc. Explainto the class that they are going to create their own picture of Rapunzel, but one which will move!

Show the children the selection of books with moving parts and talk about the effect the moving parts have on the story. Show any different types of mechanism, e.g. slider, lever, wheel, and discuss how they can be used. Write these key words up on the board, along with other important vocabulary, e.g. turn, bend, slide, join, tools, materials. Then suggest well-known stories to the children and ask them to suggest how moving parts could be used, e.g. Pinocchio and

his extending nose (slider), Snow White eating the poisoned apple (lever). End this discussion with 'Rapunzel' and introduce the picture they will be making – Rapunzel letting down her long hair. Show a prepared example (see Teachers' Notes) and ask the children to say how they think it was made – what was done first etc. Write this order of work up on the board so it is there for all to see and use during their work. This discussion will lead into them completing their Activity sheets.

 ## Group activities

Split into groups and complete the Activity sheets.

 ## Using the Activity sheets

Activity sheet 1: this is for pupils who will need regular support and supervision when completing both the written and practical work.

Activity sheet 2: this is for pupils who will need some support when completing both the written and practical work.

Activity sheet 3: this is for pupils who are able to independently work through the written and practical work, asking for support only if needed.

 ## Developing their designs

Once the children have completed their Activity sheets they can prepare to make their pictures. Remind them of the Order of Work which the class came up with together in the whole class discussion (have this on view) and then support the children in the task as they work. Build in a couple of quality-control points, where difficulties and solutions are discussed and the children are kept focussed on producing a product that will move well.

Rapunzel's Hair

 Plenary session

Gather the class together and explain now that they have made their moving pictures it is important to think about what they could have done better.

Talk through a couple of examples together and then ask the class to make pairs. In their pairs, complete their Evaluation sheet about both their picture and their partner's.

Come back together and ask pairs to read out their ideas. Extend the discussion by asking such questions as: What did you enjoy the most? What did you find easy? What did you find hard? Did you help anyone? Could anything have been done better – how? Together, agree on a statement of improvement for their next DT project.

The class can now work on writing a simple non-chronological report of their project.

 Follow-up ideas for Literacy

✦ Make a word bank of all the specific language they used during the task, e.g. hinge, bend, join
✦ Sequence the story of Rapunzel

 Follow-up ideas for DT

✦ Have an open design session with a theme, e.g animals, and challenge the children to explore different materials to design their own moving animal. Encourage the use of specific vocabulary and evaluation
✦ Explore and compare another type of moving mechanism with the children

✦ Rapunzel's Hair ✦

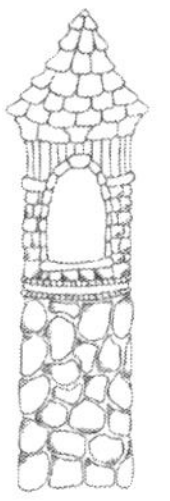

✦ You are going to make a moving picture using a SLIDER mechanism.
✦ Look at the pictures below and colour in what you think you will need.

✦ Look at the picture below and join the labels to the correct places.
  Draw Rapunzel at the top of the tower, with her long hair hanging down

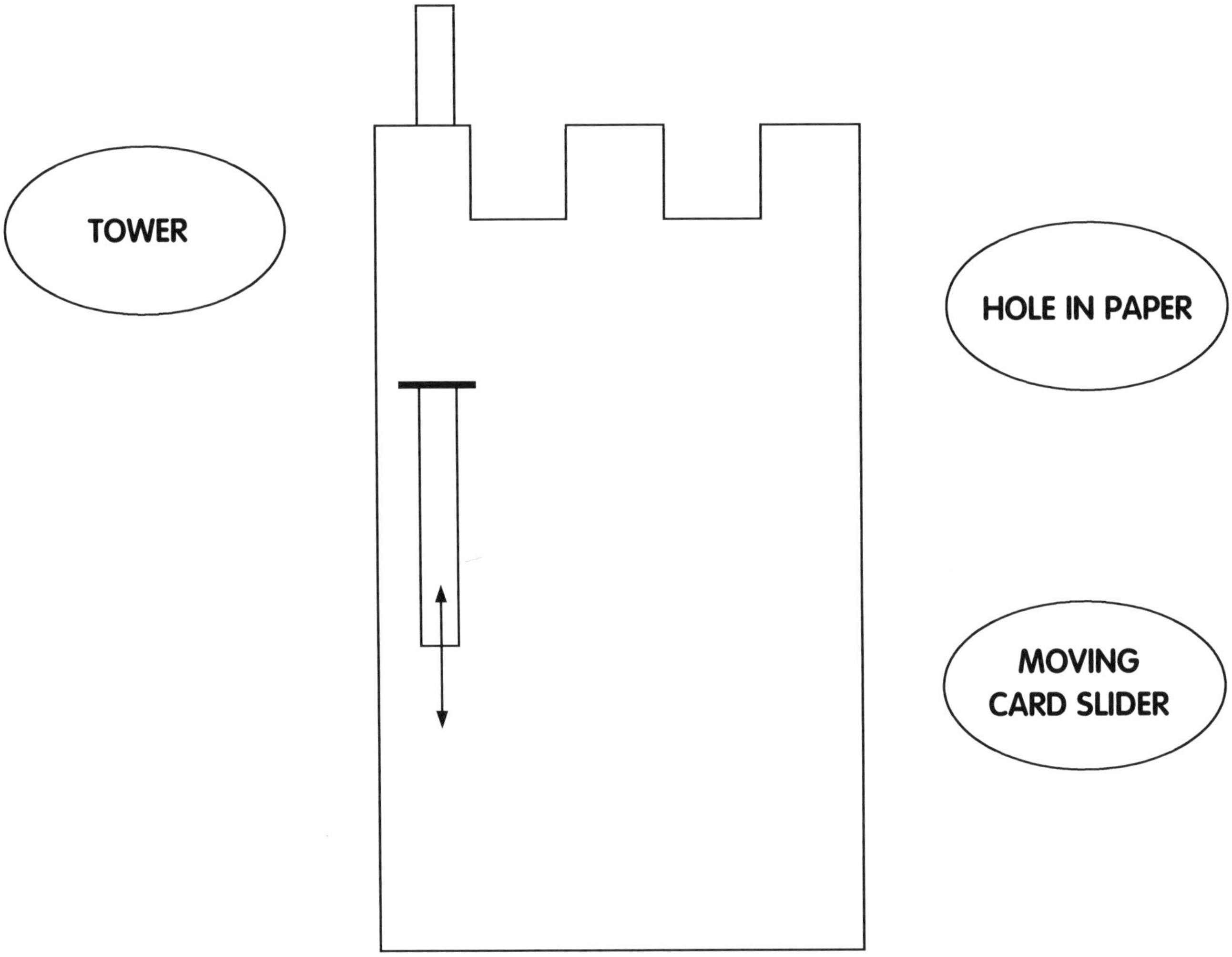

TOWER

HOLE IN PAPER

**MOVING
CARD SLIDER**

✦ Rapunzel's Hair ✦

✦ Look at the picture below. Draw Rapunzel at the top of the tower, with her long hair hanging down. Label the important parts of the picture.

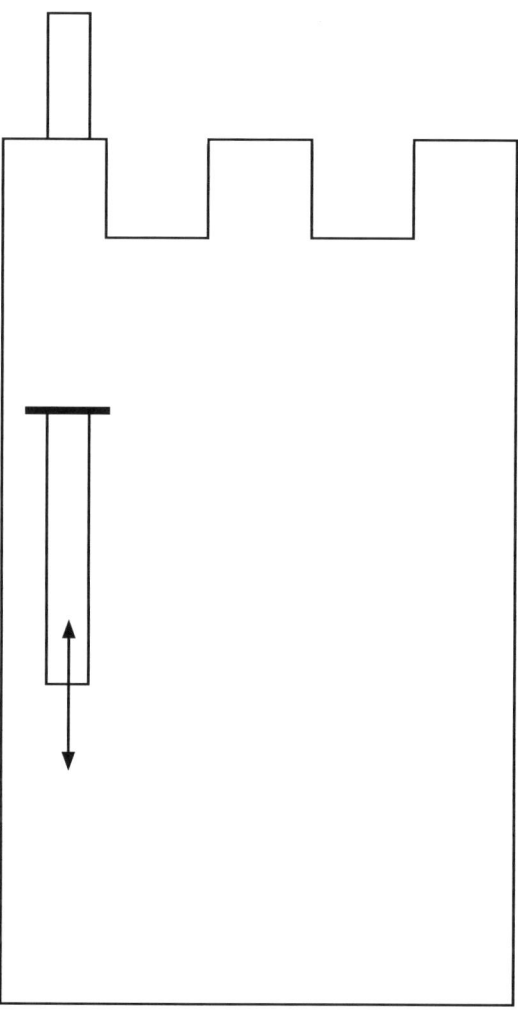

✦ Circle the things you think you will need to make your moving picture:

Scissors   Thick card   Glue   Drawing Pin   Ruler

Master sheet   Colouring pens   Pencil   Pipe cleaner

✦ What will you need to do very carefully when making your picture? ....

....

✦ Rapunzel's Hair ✦

✦ You are going to make a moving picture using a .... mechanism.

✦ The part that will move will be ....

✦ Look at the picture below. Draw Rapunzel at the top of the tower, with her long hair hanging down. Label the important parts of the picture.

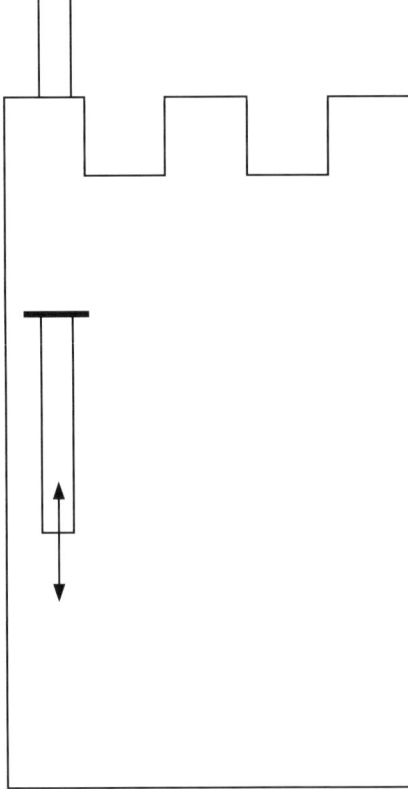

✦ Make a list of everything you think you will need to make your moving picture:

....

....

....

....

....

Literacy through design & technology

Evaluation

✦ With a friend, think about your moving pictures and answer these questions.

	My picture	.... s Picture
Does it move?		
How does it move?		
What works well?		
Could anything be improved?		

✦ The master worksheet can be photocopied on to card for each individual child to use as the starting point for their moving picture.

✦ It is important that the measurements of the card slider that the children make are

a) Wide enough to be able to fit through the cut slit

b) Long enough to be able to lengthen Rapunzel's hair to the ground

✦ Other successful moving picture ideas for

a) slider movements: Pinocchio's nose, Jack's beanstalk, troll jumping on to the Billy Goat's bridge

b) lever movements: Snow White holding an apple, then eating it, a fairy's wings moving up and down, a pirate moving a ship's wheel from side to side

c) wheel movement: The cow jumping over the moon,

Master Sheet

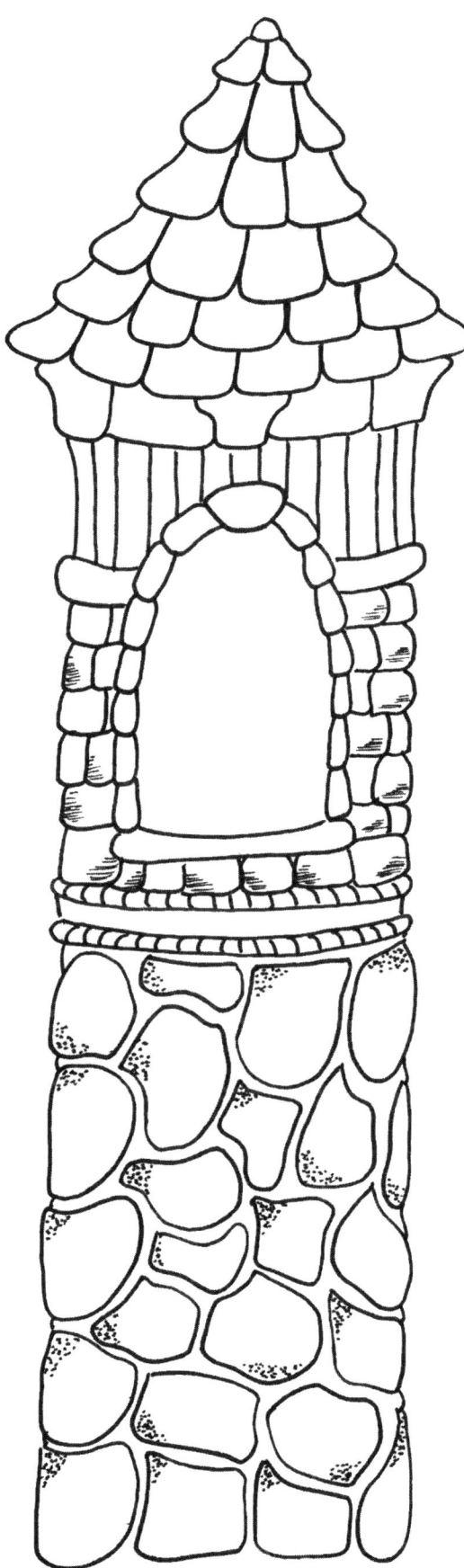

Chapter 3

To The Playground!

 Aim
✦ To use observations of existing playground structures to design a structure of their own

 Time Allocation
✦ 5 x 60 minute sessions, plus visit to playground

 Literacy Objectives

✦ To write simple questions for a survey
✦ To sort a collection of words alphabetically

 DT Objectives

✦ To recognise how structures have been designed to meet the needs of children
✦ To make a model play structure which reflects their ideas
✦ To evaluate their model, identifying strengths and areas requiring change

 Resources

✦ To The Playground! worksheets. These can later be collated into a booklet, with a front cover designed by the children.
✦ Books/photos/slides of playground equipment
✦ Construction kits, sheet materials (card, plastic), reclaimed materials (egg boxes, cotton reels, match boxes)
✦ Other materials (glue, string, Sellotape, plasticine, paint, buttons, felt pens, scissors, stapler, paper clips etc)

 Starting point: Whole Class

Begin with a class discussion about structures and equipment in a playground, writing up the children's ideas along the way. What different structures can you find in a playground? What is the purpose of each one? What skills does each structure require/develop? How do the structures appeal to children? What are your favourites and why? Which structures do you not use and why? How do you think playgrounds have changed over the years? Before there were playgrounds, what do you think children used to play outside? Images on the whiteboard or on postcards can be used to stimulate and respond to the children's thinking.

If possible, follow up this initial discussion with a visit to a local playground, or the equipment in the school playground. The children will need to observe, record and evaluate the structures there (see 'A Visit To The Playground') in addition to using them as well!

Back in school let the children report back on their observations from their visit. Explain their next task – to discover which types of playground equipment are popular in the class as a whole. They will need to write a short survey, so that they can discover what structures are popular. Recap on how to write simple questions and how to set out a survey so that it is easy to record and evaluate the responses. Let the children break in to groups to write their questions (they could be typed up in an ICT session) and then allow time for the surveys to be conducted. The results could be used in a later maths session to produce a pictogram of the class's favourite playground equipment, adding to a display wall of the class's DT task.

Together explain to the class that they are now going to think about designing a playground structure. The scale of the models will be so that they are the appropriate size for play people. The structures must be stable. They will need to think about the materials they will want to use and how to join the parts together securely. Write up key words the children will be using throughout the task e.g. join, material, cut, stable etc. Remind the class of points they learnt from their past DT projects.

 Group activities

In differentiated groups, before the children begin their own designs, show them an example of the type of structure they will be making (see Teachers' Notes). By asking the children to say how they think the structures were made, in what order things were done, how it has been joined well, why is it so stable etc. They will begin to consider those aspects that will be important in their own designs. Encourage the children to come up with their own designs, and not necessarily copy the example.

 Using the Activity sheets

Activity sheet 1: this is for pupils who will need support throughout the task, in writing out their design ideas, formulating an order of work, manipulating tools and materials successfully to build their structure.

Literacy through design & technology

To The Playground!

Activity sheet 2: this is for pupils who will require some support to transform their design ideas into an actual model, successfully combining materials and components together to construct a stable model.

Activity sheet 3: this is for pupils who may require some initial support in the task and can then work independently on the design and construction of their model, which uses a variety of components and incorporates some movement.

 ## Developing their designs

Before starting practical work, remind the children again of work done in previous tasks, and points they wanted to remember in order to help them improve in their work.

The children can now begin to build their model using the materials they listed, following their order of work. During this part of the process, be on hand to advise with their work, help them to solve any problems and pose open-ended questions to encourage the children to develop their designs. Time deadlines will also need to be made clear, to support the children towards finishing their task in a steady, focused manner. Build in at least one quality-control point, where the class stops work and you talk about the designs – what is working successfully, what is needing adaptation etc.

 ## Plenary session

✦ Self evaluation: ask the children to now think about their model designs and to complete the Evaluation worksheet.
✦ Group evaluation: within each group, ask the children to identify one design that they felt worked very successfully and explain to the class why this was so. Then identify some points where another design could be improved.
✦ As a whole class, lead the class towards thinking about how they could build on their knowledge from this project to improve their DT skills. Discuss the following - What parts did they find easy? What was hardest and why? Was there anything they did better in this project than in others? What could have been done better and how? What did they enjoy the most about this project? The whole class can then agree on a short statement to help towards improving their work in the next task.
✦ All together brainstorm all the vocabulary the children have used in this one task – the children will enjoy seeing how many words they can think of! Use their answers to then challenge the children to sort them all into alphabetical order.

 ## Follow-up ideas for Literacy

✦ To write a report about their visit to the playground
✦ To read and write poems on the playground theme
✦ To make a 'Playground' themed list of reading words

 ## Follow-up ideas for DT

✦ To develop the task to design a piece of equipment for either a specific user (a child with disabilities or special needs) or for a specific theme (Pirate Playground)
✦ To develop the idea of playing structures by asking the children to design and build e.g. a marble run, using cardboard tubes, cardboard boxes, paper cups etc.

Literacy through design & technology

✦ Designing A Slide ✦

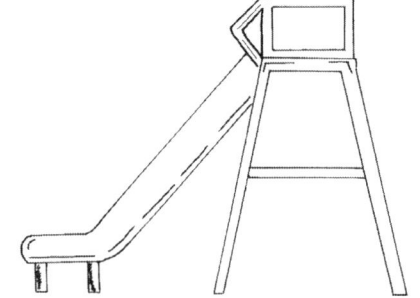

✦ **My slide will:**

> **Look good**
>
> **Stick together well**
>
> **Be steep enough for a toy person to slide down**

✦ **My slide will look like this:**

```
┌─────────────────────────────────────────────────────────────┐
│                                                             │
│                                                             │
│                                                             │
│                                                             │
│                                                             │
│                                                             │
│                                                             │
│                                                             │
│                                                             │
│                                                             │
│                                                             │
└─────────────────────────────────────────────────────────────┘
```

✦ **Neatly label all the different parts of your slide**

I will need:	**To make my slide ...**
....	**First I will** ....
....	.... **Next I will** ....
....	...............................**Then I will** ....
....	....
....	**After that I will** ....

Activity 2

Name _____

✦ My wall rope will:
    Look good
    Hold together well
    Be tight enough to support a toy person

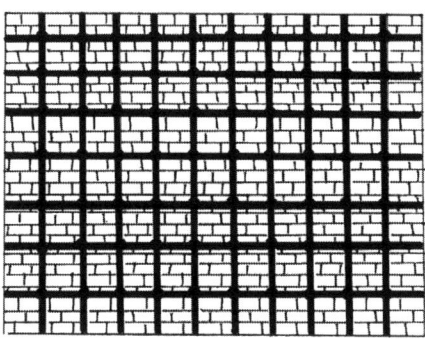

✦ My wall rope will look like this:

✦ Neatly label all the different parts of your wall rope equipment

I will need:	To make my wall rope ...
....	First I will ....
....	.... Next I will ....
....	....Then I will ....
....	....
....	After that I will ....

Literacy through design & technology

✦ Designing a Swing ✦

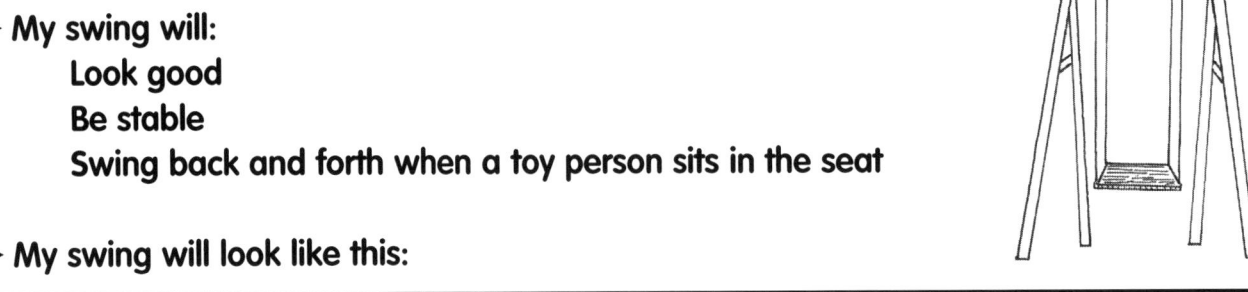

✦ My swing will:
     Look good
     Be stable
     Swing back and forth when a toy person sits in the seat

✦ My swing will look like this:

```
┌─────────────────────────────────────────────────────────┐
│                                                           │
│                                                           │
│                                                           │
│                                                           │
│                                                           │
│                                                           │
│                                                           │
│                                                           │
│                                                           │
│                                                           │
│                                                           │
└─────────────────────────────────────────────────────────┘
```

✦ Neatly label all the different parts of your swing

I will need:

....

....

....

....

....

To make my swing ...

First I will ....

.... Next I will ....

................................Then I will ....

....

After that I will ....

Evaluation

Evaluating my structure	✔	✗
Does my structure look good?		
Is my structure well-joined?		
Is my structure stable?		
Does my structure serve its purpose? (e.g. does my swing move?) Could anything be improved?		

What was difficult about this task? ....
....

What was easy about this task? ....
....

Would you do anything differently another time? ....
....

Name _____

On .... our class visited the

playground at ....

Which pieces of equipment did you see at the playground?

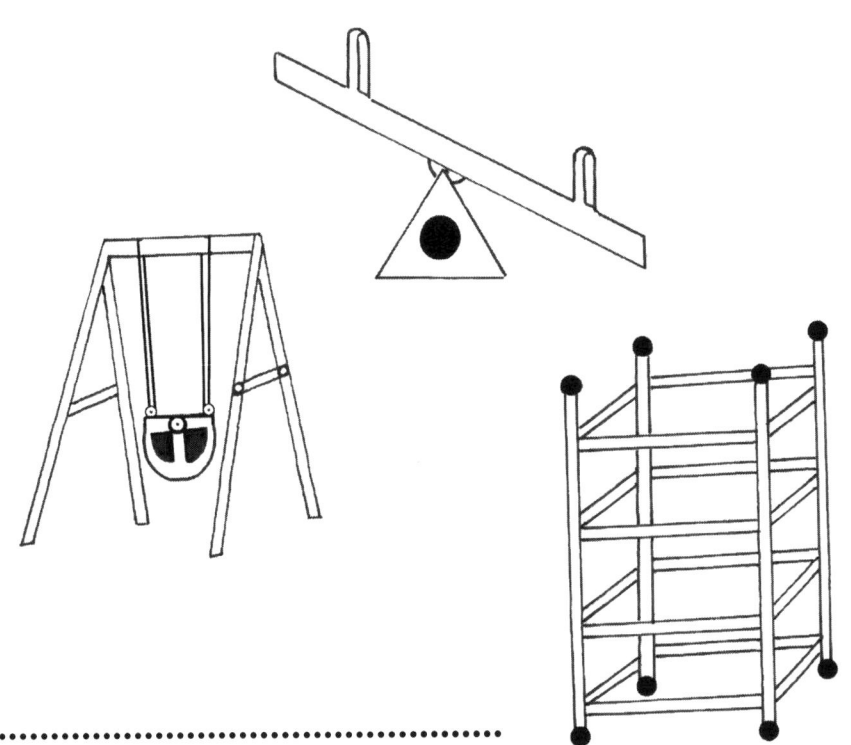

- Slide
- Baby Swings
- Child swings
- Roundabout
- See-saw
- Rope climbing
- Zip wire
- Obstacle course
- Playhouse
- Hopscotch
- Other..

Which piece of equipment needs you to balance? How?

....

....

Which piece of equipment needs you to climb? How?

....

....

Name _____

✦ Draw a piece of equipment you enjoy. Label the different parts. Label the different materials used.

```
┌──────────────────────────────────────────────────┐
│                                                    │
│                                                    │
│                                                    │
│                                                    │
│                                                    │
│                                                    │
│                                                    │
│                                                    │
│                                                    │
│                                                    │
└──────────────────────────────────────────────────┘
```

I like the ....

because ....

....

✦ Is the playground a good place for children to play?

	Yes	No
Does the playground look fun?		
Is the playground safe?		
Is there a good range of equipment?		
Could a lot of children play here safely?		

These three tasks are differentiated, although can also be adapted to be made more challenging/easier to complete.

The Slide

What you need:

+ Match boxes
+ Thick card
+ Pipe cleaner
+ Glue
+ Paint

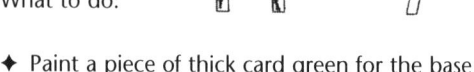

What to do:

+ Paint a piece of thick card green for the base
+ Stick matchboxes of different sizes on top of each other to create the steps of the slide. Paint.
+ Use a piece of thick card for the slide. Score, bend and stick to the top step. Stick to the base board at the bottom.
+ Cut a pipe cleaner in half and fix to the top step to form two arched handles

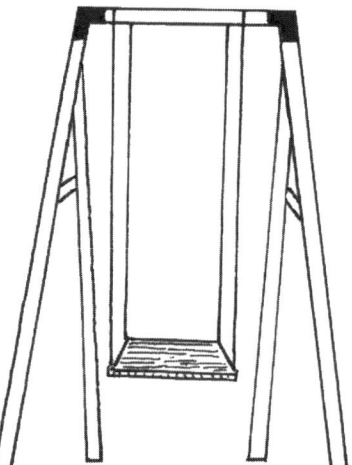

The Swing

What you need:
+ Thick card
+ Construction rods/straws for the frame
+ Frame connectors e.g. pipe cleaners
+ Thread/string
+ Egg box unit
+ Thin wire
+ Glue
+ Paint

What to do:

+ Paint a piece of thick card green for the base
+ Construct the frame, with two wires hoops on the upper crossbar and make stable by attaching card corners
+ Paint an egg box unit for the swing's seat
+ Attach a thread/string of equal length to the seat and fix to each wire loop.

The Wall Rope

What you need:

+ Thick card
+ Cardboard box
+ Thread/string
+ Glue
+ Paint

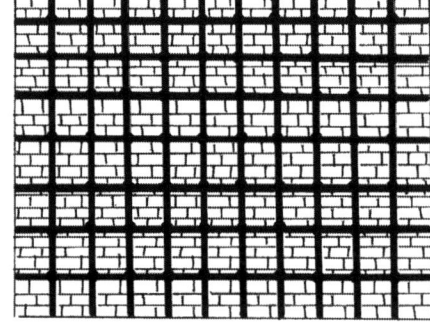

What to do:

+ Paint a piece of thick card green for the base
+ Paint a cardboard box to look like a wall and stick to the base board
+ Attach the outer pieces of string firmly to the base board and wall by threading through and tying taut.
+ Use string/thread to weave across a secure 'web' effect

Pull-along Toys

 Aim

+ To design and make a pull-along toy

 Time Allocation

+ 4 x 60 minute sessions

 Literacy Objectives

+ To use clear labels on a diagram for explanation
+ To sequence a series of instructions

 DT Objectives

+ To understand how wheels work and how the shape of the wheel affects the movement
+ To develop specific vocabulary to discuss wheels and axles with their peers

 Resources

+ Pull-along Toys worksheets. These can be collated into a workbook with a front cover, designed by the children.
+ Resources for the building of the toys: polystyrene balls of different sizes, thread, string, wooden dowels, thick cardboard/wooden wheels, rubber washers, strong glue, cardboard box selection, scissors, paints, felt.
+ A demonstration model of each type of toy/one model with interchangeable wheels
+ Selection of pull-along toys – demonstrating different types of movement if possible.

 Starting point: Whole Class

Gather the class together and examine the selection of pull-along toys collected together. By introducing the pull-along toy as one which has been around since Ancient Egyptian times, ask the children to consider why they think it has always been such a popular toy for young children. Did any of them have such a toy? What do such toys need in order to move? Draw the children's attention to any of the toys which display a different type of rolling action e.g. straight along, up and down, wibbly-wobbly. Why does this happen?

Tell the children that they are each going to design and make their own pull-along toy. They are first going to investigate some prepared models, to understand how they are constructed. (see Teachers' Notes)

First show an example of a pull-along toy which moves in a straight and steady line. Draw a simple diagram of it on the board and together label the main parts (body of the toy, wheels, axle, rubber washer, string). Explain to the children how the circular wheels give the toy a smooth and straight movement. How are the wheels kept on?

Now introduce the second toy - or show the second set of wheels, oval in shape – and ask the children to suggest how these wheels might change the way the toy moves if they are both mounted in the same direction. Show and explain how these give an up and down movement to the toy. Does the size of the wheel make any difference?

Now introduce the third toy, which has oval shaped wheels mounted in opposite directions and ask the children to describe how this toy will move when pulled along. Show and explain how these wheels give a wibbly-wobbly movement.

 Group activities

Break the children up into three differentiated groups to design and make their pull-along toys.

 Using the Activity sheets

Activity sheet 1: this is for pupils who will need a large degree of support with the design development and construction of the task, but can decorate it independently.

Activity sheet 2: this is for pupils who will need some guidance with their design development and construction, and who can decorate independently.

Pull-along Toys

Activity sheet 3: this is for pupils who are able to develop their designs independently and complete the making of their toy by themselves, after initial instruction.

Developing their designs

✦ Before starting any practical work, remind the children of the work done in previous tasks, and points they wanted to remember in order to help them improve in their work

✦ Following on from the sequencing activity, the children can then use the ordered instructions to guide them through the construction of their vehicles. Ask them first to complete the 'Design' sheet. Build in a check-in point along the way, where the children stop and show their group how they are getting on. Encourage them to reflect on the process so far,

what has been easy and what has been difficult. Are there problems that need solutions? Can the children come up with suggestions?

Plenary session

✦ Self evaluation: ask the children to now think about their toy designs and to complete the Evaluation worksheet.

✦ Group evaluation: as a whole class, lead the class towards thinking about how they could build on their knowledge from this project to improve their DT skills. Discuss the following - What parts did they find easy? What was hardest and why? Was there anything they did better in this project than in others? What could have been done better and how? What did they enjoy the most about this project? The whole class can then agree on a short statement to help towards improving their work in the next task.

Follow-up ideas for Literacy

✦ To write a story, including their toy
✦ To write a questionnaire for a younger class about toy design
✦ To prepare a short presentation to give to a younger class, describing how they made their toys and the easy and difficult things about it
✦ To read and sequence other sets of instructions

Follow-up ideas for DT

✦ Explore different joining techniques – how to join axles to wheels
✦ To make simple drawings of different vehicles and to name and label the parts
✦ Explore the effect of changing the position of the hole in a wheel

Pull-along Toys

✦ Look at the instructions below. Cut them out and stick them on to another piece of paper in the correct order.

From thick cardboard, draw and cut 4 round wheels of the same size.

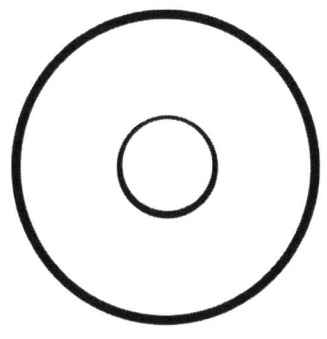

Paint the outside of your vehicle so it resembles your design e.g. fire engine, sports car. Leave to dry.

Push two wooden dowels through the box and attach the 4 wheels and 4 washers.

Attach a thick string firmly to the front of the toy, so it can be pulled along.

Make the shape of your toy vehicle by sticking a bigger and smaller cardboard box together.

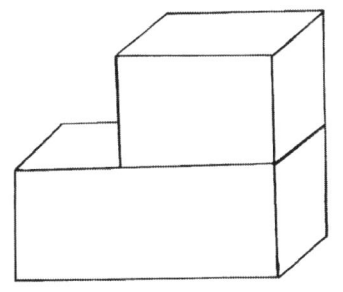

How To Make A Pull-Along Toy

Pull-along Toys

✦ Look at the instructions below. Cut them out and stick them on to another piece of paper in the correct order.

From thick cardboard, draw and cut 4 oval wheels of the same size.

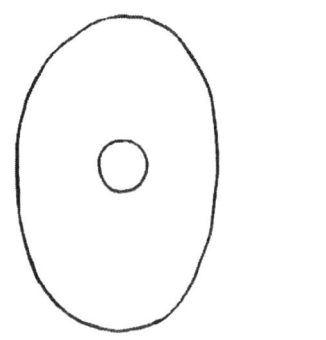

Cut out several copies of your animal shape from thick cardboard and stick them together. When dry, paint all around the outside.

Make two holes and push two wooden dowels through the animal and then attach the 4 wheels and 4 washers.

Attach a thick string firmly to the front of the toy, so it can be pulled along.

Draw the outline of the animal you are going to use for your toy.

How To Make A Pull-Along Toy

Think now about how big your wheels and your two wooden dowels need to be!

✦ Pull-along Toys ✦

✦ Look at the instructions below. Cut them out and stick them on to another piece of paper in the correct order.

From thick cardboard, draw and cut 2 oval wheels of the same size.

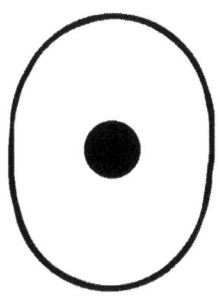

Take 1 large polystyrene ball and paint it to look like the snake's head.

Make a hole in the snake's head for the wooden dowel to go through. Attach both wheels and washers, with the wheels turned in opposite directions!

Take the snake's body and thread the long end through the snake's head to attach them together. Tie the thread and the pull string together firmly and hide the join by sticking over a felt snake's tongue.

Take a long piece of thread and string together a collection of smaller polystyrene balls. This is the body of your snake. Paint the balls and leave them to dry.

Knot the string at one end and leave a long length of thread at the other.

How To Make A Pull-Along Toy

Think now about how big your wheels and your wooden dowel need to be!

✦ Pull-along Toys ✦

✦ **Draw a diagram of your moving toy and label it clearly:**

The movement will be ....

....

✦ **The materials I will need to make my toy are:**

Evaluation

Evaluating My Toy	✔	✘
Does my pull-along toy look like I intended it to?		
Is my toy joined together well?		
Does my toy move smoothly when pulled along?		
Does my toy move as I intended it to?		
Is my toy well finished?		

What was difficult about this task? ....

....

What was easy about this task? ....

....

Would you do anything differently another time? ....

....

◆ Straight Movement Toy

The wheels for this model are circular, with the axle fixed through the centre. The car is made out of cardboard boxes, painted and decorated to the desired design e.g. fire engine, racing car. The two axles are fixed through the vehicle base by making holes in the lower side of the box.

◆ Up and Down Movement Toy

The wheels for this model are oval in shape, with the axle fixed through the centre. The animal-shaped toy is made from various pieces of thick cardboard stuck together, all cut using the same template. The two axles are fixed thorough the animal by making holes along the lower side.

◆ Wibbly-Wobbly Movement Toy

The wheels for this model are oval in shape, with the axle fixed through the centre. The wheels are fixed on the axle in opposite directions, to create a wobbly movement when pulled along. The axle is fixed to the snake by piercing a hole through the snake's head (a large polystyrene ball). Beads are threaded on to a length of string to create the snake's body, which in turn is threaded through the polystyrene snake's head.

Literacy through design & technology

KS1: Y1–2

Performing Puppets

 Aim

+ To design and make a puppet for a group performance

 Time Allocation

+ 5 x 60 minute sessions

 Literacy Objectives

+ To read and follow simple written instructions
+ To talk and write about the activity undertaken
+ To write a short, collaborative play script

 DT Objectives

+ To plan resources required
+ To produce a simple annotated drawing of what has been made
+ To join materials using different techniques

 Resources

+ Performing Puppets worksheets. These can be collated into a workbook with a front cover, designed by the children.
+ Collection of materials for puppet making e.g. socks, felt, calico, buttons, string, wool, eyes, thick card etc.
+ Puppets as demonstration models – sock puppets/finger puppets
+ Two stimulus stories – for example, Three Billy Goats Gruff and Little Red Riding Hood
+ Digital camera

 Starting point: Whole Class

Start by giving the class a short puppet performance with two/three puppets, either a self-contained short story or an extract from a well-known tale. Afterwards talk to the class about the differences between listening to a story being read to them and watching a performance. Examine the puppets you have used and ask the children to put forward ideas as to how they have been made. What materials have been used? In what order have things been completed? What do they like about the puppets? Could the puppets be improved? Suggest different story characters and ask the children for ideas of how to make puppets for them.

Explain the task to the class. They are going to work in small groups to produce some puppets from a well-known tale, which they are then going to use in a short performance. Each group will need to plan the resources they will need, decide upon an order in which to make them, and use a variety of joining techniques to successfully make their puppets. Show the methods used with the demonstration puppets to join materials together – e.g. sewing, Velcro, gluing, threading. The groups will also need to work together to write a short script to use for their performance.

 Group activities

Break off into groups to begin work on the puppets. Both groups will need an adult initially to read them the story from which their puppets will be made. For each story, two differentiated groups can be made – the sock puppets are more suited to the more independent workers, the finger puppets for those children needing more support.

 Using the Activity sheets

Activity sheet 1: this is for pupils who will need support with reading and writing, and greater direction with the more practical activities e.g. cutting material to size, threading needles.

Activity sheet 2: this is for pupils who are able to work more independently on the reading and writing tasks, and need only initial support/supervision during the practical stages.

 Developing their designs

+ Before starting practical work, remind the children of work done in previous tasks, and points they wanted to remember in order to help them improve in their work.
+ Following the instructions on their Activity sheets, guide and oversee the children through the making of their

Performing Puppets

puppets. Be on hand to advise where needed, and give time deadlines to ensure work is completed in a steady, focused way. Remind the children to use a variety of joining techniques, and upon completion ask the children to produce a simple annotated diagram of their product.

✦ Taking digital photographs at various stages of construction means that you can later ask the children to produce a booklet/worksheet describing the puppet-making process.

✦ Once the puppets have been made, groups can work together – with support where needed – to produce short play scripts, which can then be used in short performances by the children.

 Plenary session

✦ Self evaluation: ask the children to now think about their puppet designs and to complete the Evaluation worksheet.

✦ Group evaluation: as a whole class, lead the class towards thinking about how they could build on their knowledge from this project to improve their DT skills. Discuss the following - What parts did they find easy? What was hardest and why? Was there anything they did better in this project than in others? What could have been done better and how? What did they enjoy the most about this project? The whole class can then agree on a short statement to help towards improving their work in the next task.

 Follow-up ideas for Literacy

✦ Produce comic strips with speech bubbles, based upon the stories used in the puppet performances
✦ Produce leaflets, advertising the puppet shows
✦ Write character studies for each of the main characters in the stories used in the puppet performances

 Follow-up ideas for DT

✦ Extend the experience of making puppets to include making a puppet controlled by other means, e.g. rods or strings
✦ Create 'Resource Lists' needed to make a selection of puppets, as suggested by the teacher
✦ Evaluate the pros and cons of different types of joining techniques

✦ Billy Goats Gruff ✦

✦ My finger puppet is going to be ...

First goat

Second goat

Third goat

Troll

✦ Draw and colour a picture of what your finger puppet will look like ...

✦ To make my finger puppet I will need ...

Instructions for Making My Finger Puppet

1. For your puppet, first cut out two felt rectangles (with arms, if you wish) to serve as the front and back. The rectangles should allow room for the two pieces to be joined at the sides.
2. Cut pieces of felt for the features e.g. nose, ears, horns and stick on.
3. Stitch two small crosses for the eyes.
4. The front and back panels can be stuck together with glue, or with stitches.

✦ Little Red Riding Hood ✦

✦ **My finger puppet is going to be ...**

Little Red Riding Hood

The Wolf

Grandmother

Woodcutter

✦ **Draw and colour a picture of what your finger puppet will look like ...**

✦ **To make my finger puppet I will need ...**

Instructions for Making My Finger Puppet

1. For your puppet, first cut out two felt rectangles (with arms, if you wish) to serve as the front and back. The rectangles should allow room for the two pieces to be joined at the sides.

2. Cut pieces of felt for the features e.g. nose, ears, clothes and stick on.

3. Stitch two small crosses for the eyes.

4. The front and back panels can be stuck together with glue, or with stitches

◆ Billy Goats Gruff ◆

✦ **My sock puppet is going to be the ...**

First Goat

Second Goat

Third Goat

Troll

✦ **Draw and colour a picture of what your sock puppet will look like ...**

✦ **To make my sock puppet I will need ...**

Instructions for Making My Sock Puppet

1. Put your sock on your hand, so that your fingers are in the toe end and the back of your hand is where the heel should be.
2. Make a cut in to the sock, between your fingers and thumb.
3. Now cut out two oval templates, both measuring approx. 7cm by 12cm. One should be cut out of material, and the other from thick cardboard.
4. Glue the fabric to the cardboard and fold carefully in half, the short way.
5. Sew the folded oval into your sock. Now your puppet can talk!
6. Now decorate your sock puppet with eyes, hair, ears, clothes etc to make them into your special character!

Literacy through design & technology

◆ Little Red Riding Hood ◆

◆ My sock puppet is going to be ...

Little Red Riding Hood

The Wolf

Grandmother

Woodcutter

◆ Draw and colour a picture of what your sock puppet will look like ...

◆ To make my sock puppet I will need ...

Instructions for Making My Sock Puppet

1. Put your sock on your hand, so that your fingers are in the toe end and the back of your hand is where the heel should be.
2. Make a cut in to the sock, between your fingers and thumb.
3. Now cut out two oval templates, both measuring approx. 7cm by 12cm. One should be cut out of material, and the other from thick cardboard.
4. Glue the fabric to the cardboard and fold carefully in half, the short way.
5. Sew the folded oval into your sock. Now your puppet can talk!
6. Now decorate your sock puppet with eyes, hair, ears, clothes etc to make them into your special character!

Evaluation

Evaluating My Puppet	✔	✗
Does my puppet look like the character I intended it to be ?		
Is my puppet stuck together well?		
Did I use more than one joining technique when making my puppet?		
Does my puppet work well?		
Is my puppet well finished?		

What was difficult about this task? ....

....

What was easy about this task? ....

....

Would you do anything differently another time? ....

....

Joseph's Cloak

 Aim

+ To design a coat of repeating patterns

 Time Allocation

+ 6 x 60 minute sessions

 Literacy Objectives

+ To make a simple flow chart to explain a process
+ To sequence a story correctly

 DT Objectives

+ To describe how fabrics are patterned and show how the pattern can be repeated
+ To make a template to mark out the size and shape of a piece of fabric
+ To successfully join fabrics neatly

 Resources

+ Joseph's Cloak worksheets. These can be collated into a workbook with a front cover, designed by the children
+ Story of Joseph and his Cloak
+ Collection of fabric samples
+ Collection of felt (other materials could be used instead), scissors, rulers, tape measures, fabric glue, sequins
+ Dolls/bears – children's own

 Starting point: Whole Class

Read the story of Joseph to the class and discuss points of the story. Why did Joseph's father give him the coat? How was the coat special? How did Joseph help the King? Why was Benjamin thought to be a thief? Ask the class to then complete the sequencing task to check understanding of the storyline, neatly sticking and colouring in the pictures.

Gather the class together again and show them a range of different fabrics, of differing patterns. Ask the children to explain what they think each type of fabric would be suitable for. Sort the fabrics into different piles e.g. patterned v. plain, strong v. flimsy etc. Finish by sorting the repeating pattern pieces from all the others and then focus on a discussion of how a repeating pattern works. Demonstrate a few to the class. Ask the class to now think about designing their own bright repeating pattern for a coat, to be sketched on paper first and then developed in to a piece on the computer, using

a suitable graphics program. Give a quick demonstration of an idea on paper being turned into a computer-generated image, before allowing the children to work on their own, using the 'Different Types of Fabric' sheet as a starting point.

Explain to the children that they are now going to make a new 'Joseph's Cloak', to fit a doll/bear brought in from home. The same criteria apply – it must be bright and have a repeating pattern – and they may use their previous design should they wish to, or develop another design instead. They will measure, cut out, assemble and decorate the cloak using felt and sequins.

Demonstrate with an old T-shirt how a pattern and template needs to be larger than the person wearing it, as material is needed for joining it together etc. Lead the children towards planning their cloak along a T-shape design, and discuss what information the children will need to make sure their template is the correct size for their doll/teddy, and how exactly they can get this information – demonstrate with an example bear. Take the children through the process of creating a simply designed cloak and then act as their scribe to write the stages up in order.

 Group activities

The children break off into groups to complete the planning stage of their task, using the differentiated Activity sheets. All children will need differing amounts of help on the measuring tasks – measuring the doll/bear, measuring out the pattern etc. Formulating an order of work is differentiated in to three levels. Once completed, the order of work could be typed up on the computer as a simple flow diagram, to be used during the practical task.

 Using the Activity sheets

Activity sheet 1: this is for pupils who will need much support on the measurement tasks, and some literacy support with the order of work task.

Activity sheet 2: this is for pupils who will need guidance on the measurement tasks, with some initial support for their order of work.

Joseph's Cloak

Activity sheet 3: this is for pupils who can complete the measuring tasks largely independently, and who can write their own order of work correctly.

 ## *Developing their designs*

+ Before starting practical work, remind the children of work done in previous tasks, and points they wanted to remember in order to help them improve in their work.
+ Once the planning sheets have been completed the children can begin to design their cloak using the materials, following their order of work. During this part of the process, be on hand to advise with their work, aid with problem-solving tasks and pose open-ended questions to encourage the children to develop their designs. Time deadlines will also need to be made clear, to support the children towards finishing their task in a steady, focussed manner. Once the cloaks have been fitted, a simple braided belt could be made to hold the cloak in place.

 ## *Plenary session*

+ Self evaluation: ask the children to now think about their own coat designs and to complete the Evaluation worksheet.
+ Group evaluation: as a whole class, lead the class towards thinking about how they could build on their knowledge from this project to improve their DT skills. Discuss the following - What parts did they find easy? What was hardest and why? Was there anything they did better in this project than in others? What could have been done better and how? What did they enjoy the most about this project? The whole class can then agree on a short statement to help towards improving their work in the next task.

 ## *Follow-up ideas for Literacy*

+ To write a character profile for Joseph.
+ Display a word bank for 'Colour' Blue – navy, turquoise, duck egg. Link to colour-mixing in art.
+ To write a story called 'The Magic Cloak', e.g. where their doll/bear is the main character

---

 ## *Follow-up ideas for DT*

+ Children make individual patterned squares, which are then sewn together to form a class quilt.
+ Study a variety of coats and examine how the materials they are made from differ with each purpose e.g, raincoat, winter quilted coats, pocket cagoules.
+ Design printing blocks to use with paint, to produce repeated patterns on paper/fabric

---

✦ Joseph's Cloak ✦

✦ Look at all the pictures below. Cut them out and neatly stick them down on a new piece of paper in their correct order.

The servants hid a silver cup in Benjamin's sack.

Jacob gave his favourite son, Joseph, a coat of many colours.

Joseph told them he was their brother and the family were happy once more.

The brothers showed the father Joseph's dirty coat and told him Joseph was dead.

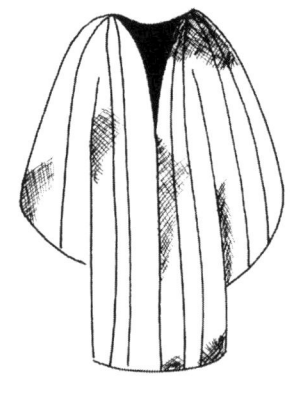

In prison, Joseph told the man he could explain to him the meaning of his dream.

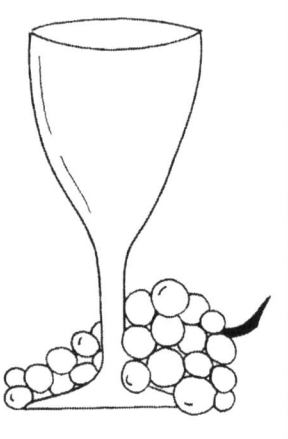

The King told Joseph about his dream, when 7 fat cows came out of the Nile.

Different Types of Fabric

✦ Draw three examples of different fabric in the boxes below and write a short description underneath each one. Colour in your drawings.

┌─────────────┐ ┌─────────────┐ ┌─────────────┐
│   │ │   │ │   │
│   │ │   │ │   │
│   │ │   │ │   │
│   │ │   │ │   │
└─────────────┘ └─────────────┘ └─────────────┘

....

....

....

....

✦ Now design a fabric of your own. It must

1)   be brightly coloured
2)   have a repeating pattern

┌───┐
│   │
│   │
│   │
│   │
│   │
│   │
└───┘

✦ Joseph's Cloak ✦

✦ Using a tape measure, measure your bear/doll and fill in the information on the sketches below. Draw and colour in your design on the cloak.

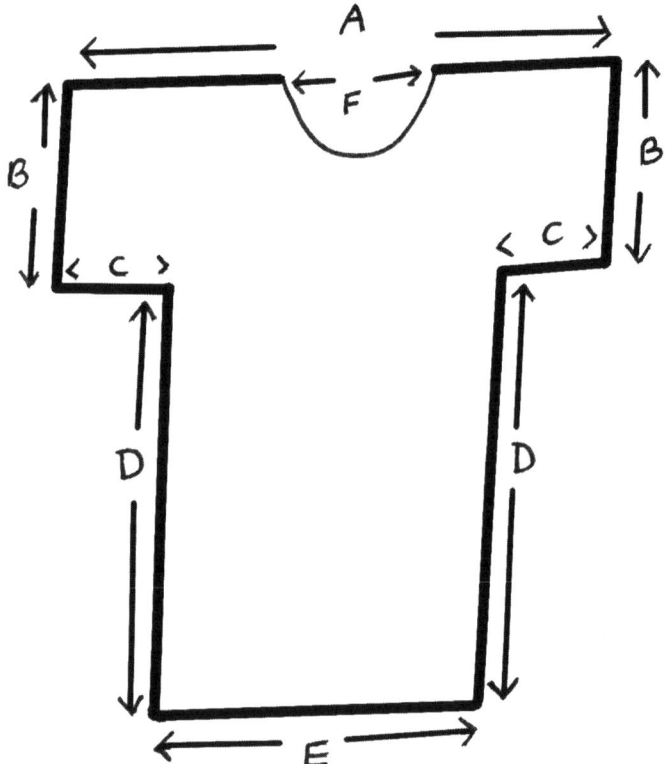

A= .... cm

B= .... cm

C= .... cm

D= .... cm

E= .... cm

F= .... cm

✦ Now draw out two paper templates to the measurements above and cut them out.

Order of work. Number the sentences 1 – 6 below, in the order in which you will work.

☐ Measure your doll/bear

☐ Carefully stick your front and back pieces together

☐ Decorate your cloak with its repeating pattern

☐ Draw and cut out a paper template for your cloak

☐ Carefully cut a hole for the head

☐ Using your paper template, cut out two pieces of fabric to make your basic cloak

◆ Joseph's Cloak ◆

✦ Using a tape measure, measure your bear/doll and fill in the information on the sketches below. Draw and colour in your design on the cloak.

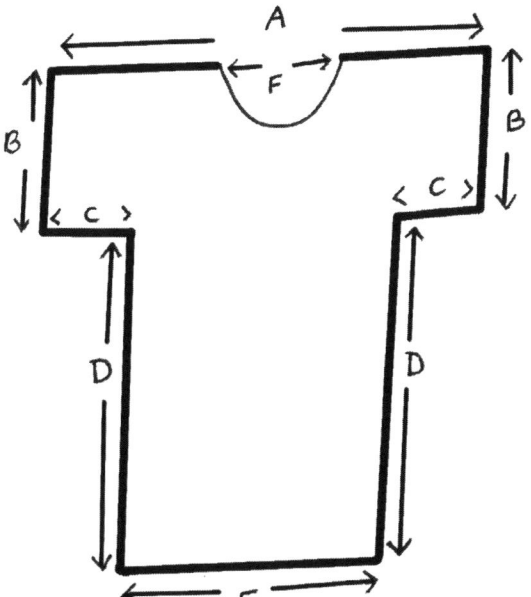

A= .... cm

B= .... cm

C= .... .cm

D= .... cm

E= .... cm

F= .... cm

✦ Now draw out two paper templates to the measurements above and cut them out.

Order of work. Complete these sentences.

1. Measure your ....

2. Draw and cut out a ....template for your cloak.

3. Using your paper template, cut out two pieces of .... to make your basic cloak.

4. Carefully stick the .... and .... together.

5. Decorate your cloak with its .... pattern.

6. Carefully cut out a hole for the ....

✦ Joseph's Cloak ✦

✦ Using a tape measure, measure your bear/doll and fill in the information on the sketches below. Draw and colour in your design on the cloak.

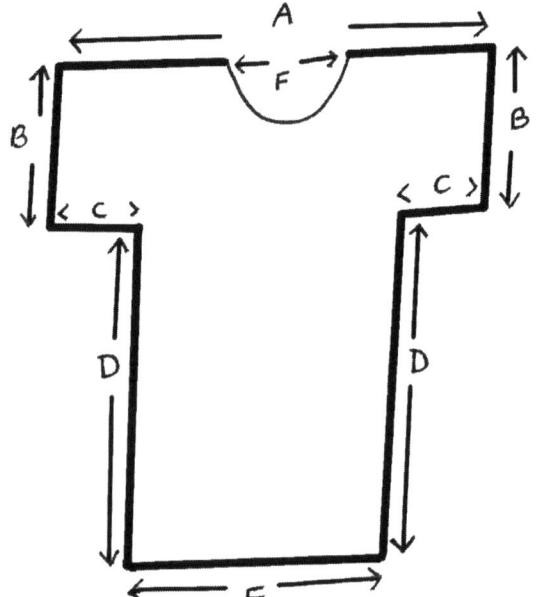

A= .... cm

B= .... cm

C= .... cm

D= .... cm

E= .... cm

F= .... cm

✦ Now draw out two paper templates to the measurements above and cut them out.

Order of work. Using the words below, write some sentences to describe your order of work to make your cloak.

cut out	paper	stick

fabric	measure	repeating	head

....

....

....

....

....

Evaluation

Evaluating My Cloak	✔	✗
Is my cloak bright?		
Is my cloak a good fit?		
Does my cloak have a repeating pattern?		
Was my cloak well cut out?		
Is my cloak stuck together well?		
Is my cloak well finished?		

What was difficult about this task? ....

....

What was easy about this task? ....

....

Would you do anything differently another time? ....

....

Notes

Notes

Notes